Cristofer Cuarezma
Carlos Mendoza

Digital Brains

Cristofer Cuarezma
Carlos Mendoza

Digital Brains

Software for Autonomous Vehicles

ScienciaScripts

Cover image: www.ingimage.com

This book is a translation from the original published under ISBN 978-613-9-46808-9.

Publisher:
Sciencia Scripts
is a trademark of
Dodo Books Indian Ocean Ltd. and OmniScriptum S.R.L publishing group

120 High Road, East Finchley, London, N2 9ED, United Kingdom
Str. Armeneasca 28/1, office 1, Chisinau MD-2012, Republic of Moldova, Europe
Managing Directors: Ieva Konstantinova, Victoria Ursu
info@omniscriptum.com

Printed at: see last page
ISBN: 978-620-8-63663-0

DEDICATION

With deep gratitude, I dedicate this work to the professors of **the Universidad Americana (UAM)**, who, with their guidance and dedication, have been beacons of wisdom and formation in this academic journey. Their teachings have not only nurtured my knowledge, but also my character and aspirations.

To my mother, **Silvia Ruiz Flores**, whose strength, unwavering love and words of encouragement have been my refuge in difficult times. Mom, your faith in me has illuminated every step I have taken.

To my father, **Fernando Cuarezma Montoya**, for being a pillar of values, discipline and motivation. Dad, your advice and constant support have been the foundation of my achievements.

To **God**, the inexhaustible source of strength, wisdom and hope, who has guided my path with divine love and made possible what seemed unattainable.

Finally, to all the people who, with their unconditional support and trust in me, have contributed to make this dream come true. Your words, actions and company have left an indelible mark on my life.

This achievement is as much mine as it is yours. With all my heart, thank you!

Table of Contents

Chapter 1: Fundamentals of Autonomous Vehicles and their Software Architecture 5

Chapter 2: Technologies, Challenges and Perspectives of Autonomous Vehicles........ 17

Conclusions........ 32

REFERENCES........ 33

FOREWORD

Since time immemorial, human beings have dreamed of achieving total autonomy in their creations. Today, that dream is materializing in autonomous vehicles: machines that not only transport us, but also think, perceive and decide. But what is behind this advance that seems to be straight out of science fiction? This book, **Digital Brains: Software for Autonomous Vehicles**, opens a window into the fascinating world of the software that brings these innovations to life.

The purpose of this text is to invite you to explore the technological heart of a revolution that promises to redefine mobility as we know it. From the sensors that allow you to "see" the world to the algorithms that decide every maneuver, this book breaks down the intricacies of the software that turns autonomous vehicles into more than just machines.

Beyond the technical, this book raises fundamental questions: What ethical and social challenges accompany this technology? How will it change the way we live and work? What role do safety and reliability play in this equation?

Throughout its pages, **Digital Brains** seeks not only to inform, but also to inspire. It is designed for readers who not only want to understand technology, but also to reflect on its impact. Engineers, students, innovators and the curious will find here a clear and motivating guide to the world of autonomous vehicles and their software.

I invite you to embark on this journey, to be captivated by the advances that are transforming transportation and to envision a future where machines and humans work together to reach new frontiers.

INTRODUCTION

This book aims to provide a comprehensive overview of autonomous vehicles, from their technical foundations to the broader implications of their implementation. By exploring both the technological foundations and the challenges to be overcome, the reader is invited to reflect on the transformative impact this technology will have on the future of mobility and society at large. This book explores in depth the fundamentals, technologies, challenges and perspectives that underpin the development of these vehicles.

Chapter 1 introduces the conceptual and technical foundations of autonomous vehicles, starting with a definition and classification of their autonomy levels. It includes a historical overview of the key milestones in their development and highlights their growing relevance. The current and future applications and benefits of this technology are examined. In addition, the software architecture that supports it is detailed, explaining its essential components, distributed systems design and the interaction between perception, decision making and control. Finally, agile development strategies and case studies that illustrate successful architectures are explored.

Chapter 2 discusses the technologies, challenges, and prospects of autonomous vehicles, highlighting the role of sensors, data processing, and artificial intelligence for real-time decision making. It also addresses safety, regulatory challenges, and social, economic, ethical, and environmental impacts of their adoption.

Chapter 1: Fundamentals of Autonomous Vehicles and their Software Architecture

Introduction to Autonomous Vehicles

In the last decade, technological advances have significantly transformed our interaction with the environment, with the development of autonomous vehicles standing out among them. These innovations not only promise to revolutionize transportation, but also to redefine key aspects of mobility, road safety and energy efficiency.

Definition and concept of autonomous vehicles

Autonomous vehicles, also referred to as driverless vehicles or intelligent cars, represent a disruptive innovation in the transportation industry. These cars operate without direct human intervention through the use of advanced technologies, such as sensors, cameras, radar, and artificial intelligence systems. The ability of these vehicles to interpret their environment and make real-time decisions is key to their safe and efficient operation.

Classification of Autonomy Levels

The Society of Automotive Engineers (SAE, 2018) classifies vehicle autonomy into levels 0 to 5. At level 0, the driver is in full control of the vehicle, with no automated assistance. As one progresses through the levels, the degree of human intervention decreases. Thus, at level 5, the vehicle is fully autonomous and capable of operating in all conditions without the need for human intervention. At level 1, the driver is assisted in certain tasks, such as steering or acceleration. At levels 2 and 3, vehicles can perform more complex tasks, such as lane changes and highway navigation, although they still require human supervision. At level 4, vehicles can operate without intervention in specific environments, while level 5 requires no driver at all.

The ability of autonomous vehicles to analyze environmental data and respond safely and efficiently involves identifying obstacles, interpreting traffic signs, and making decisions related to vehicle speed and direction. These innovations not only redefine the driving experience, but also pose new challenges and opportunities in transportation.

Figure **1**

Classification of autonomy levels.

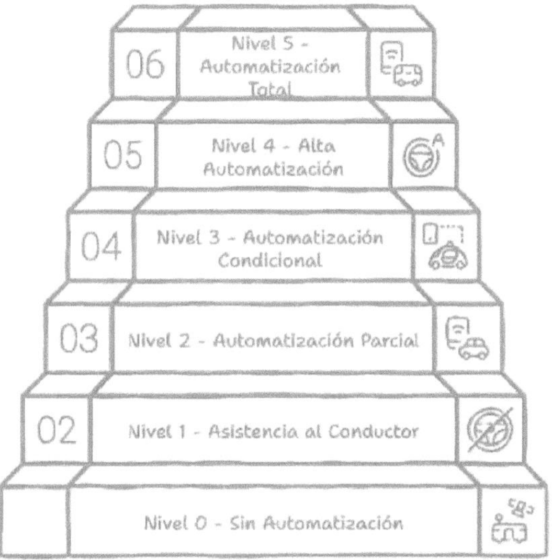

Source: Own elaboration.

History of Autonomous Vehicles

The history of autonomous vehicles dates back to rudimentary experiments in the early 20th century. One of the earliest examples was a remotely controlled vehicle by Nikola Tesla in 1898, although this was not autonomous in today's sense. In the 1980s, the development of "Navlab" by the Massachusetts Institute of Technology (MIT) marked a significant breakthrough in machine vision applied to vehicular navigation (Anderson et al., 2016). This project demonstrated that it was possible to design vehicles capable of "seeing" their environment and reacting accordingly.

A major milestone was the DARPA Grand Challenge in 2004, which challenged research teams to develop vehicles capable of driving a course without human intervention. Although many failed to complete the course, this event significantly boosted research in the field. Companies such as Google (now Waymo) and Tesla have since led the way in the development of autonomous vehicles, conducting extensive testing and offering semi-autonomous features through software updates.

Figure **2**

History of autonomous vehicles.

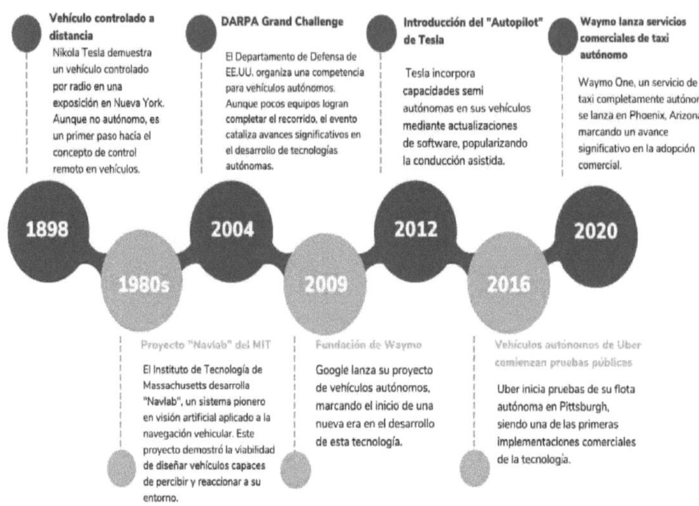

Source: Own elaboration

Importance of Autonomous Technology

Autonomous technology impacts multiple social and economic aspects. First, it is expected to improve road safety, as more than 90% of car accidents are due to human error (Litman, 2020). By eliminating the human factor, a significant reduction in traffic-related accidents and fatalities is anticipated. In addition, this technology can optimize traffic efficiency through vehicle-to-vehicle (V2V) and smart infrastructure (V2I) communication, which improves the user experience and reduces emissions generated by vehicle congestion (Fagnant & Kockelman, 2015).

In economic terms, the adoption of autonomous vehicles is expected to generate new industries related to their production and maintenance, fostering jobs in engineering and artificial intelligence. Finally, one of the most relevant aspects of autonomous technology is that it facilitates accessibility for people with disabilities or limited mobility, allowing them to move around without relying on traditional public or private transportation (Goodall, 2014).

In conclusion, autonomous technology represents a crucial

breakthrough, offering practical solutions to societal problems such as road safety, traffic efficiency and accessibility.

Applications and Benefits of Autonomous Driving

Autonomous driving is transforming the way people and goods move, offering innovative solutions to problems of safety, efficiency and accessibility. This technology not only promises to revolutionize public and private transportation, but also to facilitate new forms of logistics and medical emergencies, improving the quality of life in several areas.

This section will explore current applications of autonomous driving in different sectors, from automated cabs and buses to deliveries and emergency vehicles. In addition, the key benefits that this technology brings will be discussed, including accident reduction, traffic optimization and social inclusion through accessibility for people with reduced mobility. These advances show the potential of autonomous driving to redefine modern transportation and its impact on society.

Current and Future Applications

Current applications of autonomous vehicles are diverse. In public transportation, autonomous buses have already been implemented in several cities, improving punctuality and reducing operating costs. Autonomous cabs are also being developed, such as Waymo's, which allow users to request rides through mobile applications, offering competitive fares and accessibility.

In the automated delivery arena, companies such as Amazon are exploring the use of drones and autonomous ground vehicles to improve logistics. Autonomous emergency vehicles, such as ambulances, could improve the speed of emergency response by avoiding traffic constraints or human fatigue.

In the not-too-distant future, autonomous technologies are expected to be integrated into recreational vehicles, tractors for automated agriculture and urban logistics trucks, where drones will complement last-mile deliveries.

The future of autonomous vehicles promises to transform not only transportation, but also the structure of cities and everyday interaction.

Expected benefits

Autonomous driving is expected to bring wide-ranging benefits, such as a significant reduction in accidents, thanks to the elimination of human error (Fagnant & Kockelman, 2015). In addition, it will optimize traffic through V2V systems that will enable smoother traffic flow. Accessibility will also be improved, facilitating independent transportation for the elderly or people with reduced mobility (Goodall, 2014).

In short, autonomous vehicles represent a major breakthrough in modern transportation. Their constant development promises to improve crucial aspects of road safety and optimize urban traffic, transforming the way we interact in our cities.

Software Architecture for Autonomous Vehicles

Software is the core that brings autonomous vehicles to life, enabling them to perceive, analyze and act in real time to ensure safe and efficient driving. Its design requires a robust architecture that integrates multiple components, such as perception, decision-making and control modules, working together to interpret the environment and execute precise actions.

In this section, we will explore the essential elements of this architecture, including the importance of distributed systems for handling large volumes of data, and how interaction between modules ensures coordinated operation. We will also discuss the advantages of agile development to adapt to rapid technological change and case studies of successful architectures, such as Waymo and Tesla, that highlight innovative solutions to today's challenges.

Key Software Components

Autonomous vehicle software is based on several essential components that work together to ensure safe and efficient operation. The main ones are:

- **Perception Module**: This module collects and processes data from the environment using sensors such as cameras, Lidar and radar. Using computer vision and machine learning algorithms, it identifies objects, obstacles and traffic signs. Accuracy is key here, as any error can affect future decisions. For example, it must clearly distinguish between a pedestrian and an inanimate object, and recognize adverse weather conditions that affect visibility (Zhou et al., 2020).
- **Decision Making Module**: Once the information has been processed by the perception module, this component evaluates the possible actions to follow. It uses techniques such as Markov Decision Process (MDP) and planning algorithms to choose the best route or action according to the context. For example, if it detects a red traffic light, it must decide whether to stop or continue, depending on the presence of other vehicles or pedestrians (Bhatia et al., 2021).
- **Control Module**: This component executes system decisions using techniques such as predictive control (PCM), adjusting the vehicle's speed and direction to keep it on the desired trajectory while responding to changes in the environment (Kumar et al., 2020).

- **Communication Interface**: Communication is crucial, both between internal modules and with other vehicles or smart infrastructures (V2V and V2I). This ability to exchange information in real time allows vehicles to coordinate their movements and improve safety (Chen et al., 2019).

Each module must be designed with robustness and reliability to ensure the safety of the vehicle and its occupants. Effective integration between these modules is vital to provide a smooth and safe user experience.

Figure **2**

Key software components.

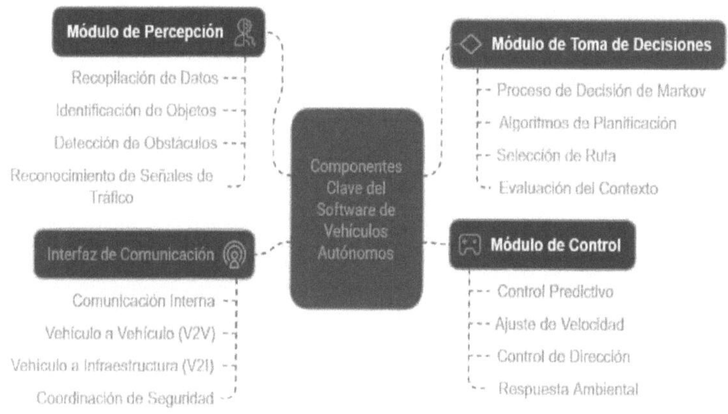

Source: Own elaboration

Distributed Systems Design

Distributed systems architecture is essential in autonomous vehicles, as it enables large volumes of data to be processed in real time. Each sensor in the vehicle generates data simultaneously, which requires a structure that handles this load without sacrificing performance.

Distributed systems allow different modules to work in parallel, improving efficiency and reducing response times. For example, while one module processes visual data, another can evaluate radar or Lidar information. In addition, this structure facilitates the implementation of redundancy strategies, where alternative modules can assume functions in case of failure, thus increasing the resilience of the system (González et al., 2021).

However, distributed systems also present challenges in synchronization and status management between modules, so it is essential to establish robust protocols to ensure that all components have access to up-to-date and consistent information.

Interaction between Perception, Decision Making and Control

The interaction between the sensing, decision making and control modules is critical to ensure safe and smooth operation.

- **Start in Perception**: The process starts with the perception module, which collects data about the environment using sensors.
- **Evaluation in Decision Making**: Next, the decision-making module evaluates possible actions based on advanced algorithms. For example, if it detects an obstacle, it must decide whether to brake, swerve or stop.
- **Execution in Control**: Once the decision is made, the control module implements the necessary actions. Effective communication between these modules is essential; any delay or error can compromise the safety of the vehicle (Gonzalez et al., 2021).

In addition, this interaction must be adaptable to sudden changes. For example, if a pedestrian appears unexpectedly, the system must react immediately to avoid an accident, which requires fast algorithms and accurate perception.

Figure **3**

Vehicle operation process.

Source: Own elaboration

Agile Development

Agile development is a popular methodology in autonomous vehicle software development due to its flexibility and iterative approach. It allows teams to adapt quickly to changes in requirements or in the market, through short development and deployment cycles. Thus, new functionalities are tested in controlled environments before their final deployment, allowing early identification and correction of errors (Larman & Basili, 2003).

One of the most salient features of agile development is its ability to adapt to changing project needs. In the context of autonomous vehicles, where technology and market requirements evolve rapidly, this flexibility is crucial. Agile teams can respond to changes in regulations, technological advances or even end-user feedback without wasting significant time or resources. This is especially important in an industry where continuous innovation can be the difference between success and failure.

Agile methodology encourages collaboration between disciplines-such as software engineers, robotics experts, and designers-which improves integration between hardware and software. This multidisciplinary collaboration is essential for the development of complex systems such as autonomous vehicles, where the interaction between physical and logical components must be seamless and efficient (Meyer et al., 2020).

The iterative approach to agile development involves dividing work into short cycles called "sprints". Each sprint culminates in a review where results are evaluated and next steps are planned. This cycle allows teams to make quick adjustments based on the feedback received, which is especially valuable when implementing new functionality or making improvements to existing systems (Schwaber & Sutherland, 2017).

The use of automated testing is another key component of agile development. These tests ensure that each new feature does not adversely affect the rest of the system. Test automation allows teams to catch bugs earlier in the development cycle, thus reducing the time and cost associated with late fixes (Crispin & Gregory, 2009). In the context of autonomous vehicles, where safety is paramount, these tests are essential to ensure that each component functions correctly under various conditions.

A common approach is to implement unit tests that verify each module individually and integration tests that ensure that all modules work together as expected. This not only improves software quality but also increases team confidence in new implementations.

Despite its advantages, the implementation of agile development in autonomous vehicles also faces significant challenges. One of the main challenges is ensuring the safety and reliability of the software. Since autonomous vehicles operate in complex environments and must interact with other road users, any error can have serious consequences (Bhatia et al., 2020). Therefore, it is essential that agile practices are complemented by rigorous safety assessments.

Case Studies of Successful Architectures

Examples of successful architectures include:

Waymo: Its modular architecture enables efficient and fast updates without disrupting operations, thanks to its use of deep neural networks for perception and MDP-based decisions. This allows it to navigate complex urban environments and adapt to unforeseen situations (Waymo LLC, 2021).

Figure **4**

Waymo.

Source: Xataka

Tesla: Tesla uses over-the-air (OTA) updates to continuously improve its autonomous software. This architecture facilitates the integration of new algorithms using data collected during daily use, allowing them to launch new features without the need to visit a workshop (Tesla Inc., 2021).

Figure **5**

Tesla.

Source: Infobae

CARLA Simulator: Used by researchers to test architectures in complex scenarios, it allows evaluating performance in different conditions without the risks of physical testing. It is an invaluable tool to validate algorithms before implementing them in real environments (Dosovitskiy et al., 2017).

Figure **6**

Carla Simulator.

Source:carlasimblog.wordpress.

These cases highlight how different architectural approaches can contribute to the overall success of effective autonomous vehicle software development and deployment. As the technology advances, it will be interesting to see how these architectures evolve to address new and emerging challenges.

Chapter 2: Technologies, Challenges and Perspectives of Autonomous Vehicles.

The creation of autonomous vehicles has revolutionized the automotive industry and the way people move around the city. Throughout this chapter, we will look in detail at the most prominent technologies that enable vehicles to operate without human intervention, and then analyze the challenges and future advantages they bring for the future. In addition, we will delve into sensors and software perception of the environment, artificial intelligence and machine learning, as well as software reliability and security to become familiar with their impact on society.

Sensors and Perception in the Environment

Sensing the environment is crucial to the safe operation of an autonomous vehicle. Vehicles use a variety of sensors to collect data about their immediate environment.

Types of Sensors Used

Autonomous vehicles use a wide range of sensors to collect data about their environment. These sensors are vital for perception and decision making. The most commonly used sensors are the following:

Cameras: They are used for object detection, traffic sign recognition and traffic light reading. A camera produces visual information that is essential for identifying nearby pedestrians and vehicles (Zhou et al., 2020).

Ultrasonic sensors: They measure the distance to surrounding spaces. Ultrasonic sensors are often used to assist in driving and when parking (Kumar et al., 2019).

Lidar: This sensor measures distance using a laser aimed at an object and a receiver to measure the time it takes for the waves to reflect, and thus detect the change in propagation speed caused by the movement of the object. It is useful for detecting obstacles at various distances and in low light (Bhatia et al., 2021).

Radar: Uses radio waves to detect data and factor measurements; this sensor is effective in bad weather, such as rain or fog, because other sensors fail (González et al., 2021).

The combination of all these sensors ensures that autonomous vehicles recognize and make decisions regarding their surroundings.

Sensory Data Processing

Sensory data processing represents a critical step in the operation of autonomous vehicles, as the data obtained through sensor collection must be analyzed in real time to enable correct and fast decisions. This stage includes a series of essential steps that convert the raw sinusoidal signals obtained by the sensors into information useful in decision making. The stages of this process are:

- **Filtering:** Removal of noise and irrelevant data through the Kalman filtering algorithm or the smoothing algorithm (Welch & Bishop, 1995).
- **Segmentation:** Identification of isolatable objects in the environment, using the region-based segmenter algorithm and clustering techniques.
- **Recognition:** Classification of detection, pedestrians and vehicles through machine learning, especially Support Vector Machines or Convolutional Neural Networks (CNN) (LeCun et al., 2015).

The techniques used in this processing are essential to ensure that the vehicle can make correct decisions.

Sensor Fusion and Information Integration

Sensor fusion is an essential process that combines information from multiple sources to improve perception of the environment. This approach allows autonomous vehicles to have a more complete and accurate view, which is crucial in complex situations.

Fusion can be performed at different levels:

- Mixing Lidar information with photographs can enhance detection in challenging situations, such as fog or rain (Doherty et al., 2017).
- The recommended systems employ methods such as the Extended Kalman Filter to compute positions more accurately by fusing inertial information with GPS (Bar-Shalom et al., 2001).

Object Detection Algorithms

Object identification algorithms are essential for the safe and efficient performance of autonomous vehicles. These algorithms facilitate the identification and categorization of elements in the environment, such as other cars, pedestrians, cyclists, and isolated obstacles.

There are various techniques employed in object identification such as computer vision based methods that employ methods such as convolutional neural networks (CNN) to process images captured by cameras (Gonzalez et al., 2021)._ These procedures have proven to be highly effective in well-lit environments, but may face challenges under unfavorable circumstances.

Lidar-based sensing that employs point clouds produced by Lidar to recognize objects through particular algorithms that study the geometry of the environment (Chen et al., 2019). This technique is particularly beneficial in circumstances where cameras may not operate properly.

Multimodal fusion is also critical, as it fuses visual data with radar or Lidar information to increase accuracy in difficult or complex situations. This method has proven to be effective in increasing the overall robustness of the system to environmental variations.

*Figure **7***

Object detection algorithms.

Source: Own elaboration

These algorithms must be robust and efficient, able to operate in real time when handling large amounts of data obtained from multiple sensors.

Perception Challenges and Proposed Solutions

Despite remarkable progress in sensory technologies, autonomous vehicles face multiple challenges related to perception:

- **Adverse weather conditions:** precipitation, snow or fog can influence camera and lidar performance. To address this issue, algorithms are being created that employ deep learning to boost recognition in challenging situations (Zhou et al., 2020). In addition, hybrid methods that fuse various types of sensors are being investigated to mitigate such constraints.

- **Complex urban environments:** Traffic intensity and the constant presence of pedestrians pose unique challenges. Sophisticated simulations are being implemented that facilitate the training of systems in various contexts prior to their actual execution (Bhatia et al., 2021). These simulations help to equip systems for unexpected circumstances through the extensive use of virtual environments in which different operating conditions can be experimented with without physical danger.
- **Visual interference:** Elements such as shadows or reflections can cause confusion in sensory systems. Preferred methods such as generative adversarial neural networks (GANs) are being investigated that could help increase recognition quality in these challenging circumstances (Gonzalez et al., 2021).

Artificial Intelligence and Machine Learning in Autonomous Vehicles

Artificial intelligence (AI) is a key component in the operation of autonomous vehicles, enabling them to learn and adapt to their environment.

AI enables autonomous vehicles to interpret large volumes of data from sensors. This data is essential for perception of the environment, including identification of obstacles, traffic signs and other vehicles. According to a recent study, the integration of advanced data engineering techniques and machine learning models significantly improves the performance and safety of autonomous vehicles by enabling more accurate and faster decisions in complex situations (Advancement of Machine Learning and its Applications, n.d.).

Introduction to Artificial Intelligence

Artificial intelligence (AI) plays an essential role in the evolution of autonomous vehicles. By implementing preferred algorithms, systems can learn to make decisions based on previous experiences, thus optimizing their performance over time . Artificial Intelligence enables vehicles not only to respond to current circumstances, but also to foresee possible future situations through predictive analytics.

- **Perception:** Improved identification and interpretation of the environment through preferred methods such as deep learning.
- **Decision making:** Allows evaluation of multiple options based on collected data to select the safest and most efficient actions.
- **Planning:** Assists in defining ideal routes taking into account elements such as traffic, weather conditions and possible barriers (Bhatia et al., 2021).

Machine Learning

Machine learning is classified into two types: supervised learning and unsupervised learning.

Supervised learning: In this methodology, a model is trained using a labeling set in which each input has a specific identified output. This procedure is beneficial for tasks such as classification and identification where high accuracy is needed (González et al., 2021). For example, supervised learning can be used to train a system to recognize pedestrians based on labeled images.

Unsupervised learning: In contrast to the supervised approach, this method deals with unlabeled sets, looking for underlying patterns or structures without prior guidance. This procedure is beneficial for clustering or dimensionality reduction, allowing the identification of hidden relationships within the data set (Zhou et al., 2020).

Figure **8**

Types of machine learning.

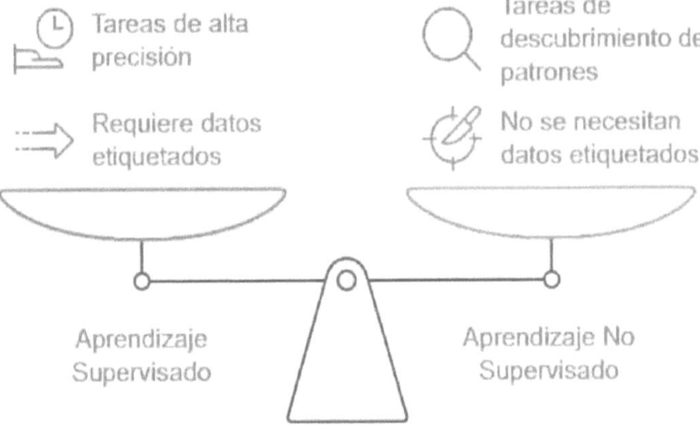

Source: Own elaboration

Both approaches have useful applications in the autonomous vehicle domain, offering flexibility depending on the particular type of problem task being addressed.

Neural Networks and Applications

Neural networks are an essential element in the deep learning employed by autonomous vehicles. These constructs are designed to replicate the operation of the human brain through linked layers that handle complex information.

The most common applications include:

- **Detection and classification:** Convolutional neural networks (CNNs) are extensively employed in the processing of images captured by cameras, facilitating the identification of objects with high accuracy (Bhatia et al., 2021).
- **Prediction:** Recurrent neural networks (RNNs) are useful for preventing future behaviors based on time sequences, such as preventing pedestrian movements or traffic modifications.

- **Adaptive control:** Neural networks can also be used to dynamically modify parameters in the control system based on varying environmental conditions (González et al., 2021).

Deep neural networks have proven to be particularly effective due to their ability to represent complex nonlinear relationships between variables (Goodfellow et al., 2016).

Recommendation and Prediction Systems

Recommender systems are useful tools that can be incorporated into autonomous vehicle software to optimize the user experience; these systems employ preferred algorithms to examine individual preferences and provide personalized recommendations on intended routes based on past behavior.

In addition, predictive systems can anticipate future events through statistical analysis of historical data collected during day-to-day vehicle operations; This involves anticipating vehicle congestion and identifying habitual patterns that may influence expected arrival time (Chen et al., 2019).

The effective implementation of recommender and predictive systems can be beneficial not only from an operational perspective, but also from a smoother and more intuitive experience for end users. This facilitates the process of making informed decisions and improves the routes chosen, thereby increasing efficiency and user satisfaction in their interaction with the system.

Incorporating predictive and recommender systems not only improves operational performance, but also changes the user experience by simplifying data-driven decision making and increasing efficiency in the use of autonomous vehicles.

Ethical Challenges in AI

As technological advancement in autonomous vehicles progresses, major ethical questions emerge regarding the widespread implementation of this new mode of future modern mobility:

Legal liability: If accidents occur with autonomous vehicles, the question arises as to who is liable-the software manufacturer itself-which poses considerable legal challenges (Bhatia et al., 2021).

Moral decisions: Algorithms need to be ready to make crucial decisions in risky circumstances; this generates ethical dilemmas about how to design these decisions correctly (Gonzalez et al., 2021). For example, if an unavoidable accident happens, should the vehicle give priority to protecting the occupants rather than protecting a larger group?

Privacy: The large-scale collection of data required to operate effectively raises concerns about how they manage that personal data and sensitive information linked to the end users involved; this requires careful handling by developers and regulators as we move towards a future in which these vehicles will be commonplace.

Software Security and Reliability

Software security and reliability represent fundamental pillars in the development of autonomous vehicles, due to the potentially serious consequences that can arise from technical failures or cyber attacks. A robust software system must ensure not only protection against external threats, but also efficient and reliable operation under various conditions.

In this section, the key principles for designing secure software will be addressed, including the validation and testing practices required to ensure its quality. In addition, approaches to risk management and incident response will be explored, as well as the regulations and standards that govern this field. Finally, future prospects for security will be discussed, highlighting emerging technologies that seek to strengthen public confidence in autonomous systems.

Software Security Concepts

Software security is essential in the development and deployment of autonomous vehicles, due to the potentially deadly consequences linked to technical errors and malicious attacks. It is essential to establish robust practices from the initial phases to the final implementation:

Security design: Including security design principles from the beginning of the development process helps to reduce vulnerabilities before real operational problems arise. This involves conducting a detailed study of the potential risks associated with each element of the system, ensuring that weaknesses are addressed before they become significant risks (Kamin, 2017).

The safe design should include:

Risk analysis: Assess potential threats and their possible impacts. This analysis must be continuous and adaptive as technologies and operating environments evolve (VeRA, 2020).

Defense in depth principles: Implement multiple layers of security to protect systems. This means that if one layer fails, others can continue to protect the system (ISO/SAE 21434, 2021).

Code review: Perform regular audits of source code to identify vulnerabilities before they are exploited. Automated tools can help detect common bugs and known vulnerabilities (Bhatia et al., 2020).

Implementing a secure design from the outset not only helps mitigate risks, but can also significantly reduce the costs associated with subsequent corrections and crisis management.

Strong encryption: Securing connections between internal modules and external connections safeguards against improper access and harmful interference during normal operation. Strong encryption is essential to protect data integrity and ensure that sensitive information is not susceptible to external attacks.

Encryption strategies should include:

Encryption in transit: Ensure that all data transmitted between vehicle components is encrypted to prevent interceptions. This is especially important in communications between the vehicle and external infrastructure (such as traffic signals or central control systems) (Kamin, 2017).

Encryption at rest: Protect data stored inside the vehicle, such as routing information or driver preferences, using strong encryption to prevent unauthorized access in case the vehicle is physically compromised.

Key management: Implement strict policies for the management of cryptographic keys used in encryption. This includes regular key rotation and immediate revocation in case of compromise (ISO/SAE 21434, 2021).

Strong encryption is an essential defense against cyber attacks, ensuring that even if an attacker gains access to a system, they cannot obtain valuable information without the proper keys.

Testing and Validation: Rigorous software validation is another critical component of ensuring security in autonomous vehicles. This includes extensive testing to identify vulnerabilities before the software is deployed in a real-world environment. Testing should cover:

Unit testing: Evaluating individual software components to ensure their correct functioning before integration.

Integration testing: ensuring that different modules of the system work correctly together. This is vital to detect problems that may not be evident when testing components separately (Schwaber & Sutherland, 2017).

Penetration testing: Simulate cyber attacks to evaluate the system's resistance to external threats. These tests help identify weaknesses in the software architecture and allow adjustments to be made before the final release (Crispin & Gregory, 2009).

Awareness and Training: Ongoing training of personnel involved in software development and maintenance is essential to maintain high security standards. Engineers must be well informed about the latest cyber threats and best practices to mitigate them (Fitzgerald & Stol, 2017).

Strategies should include:

Regular training programs: Update staff on new technologies and methods for securing software.

Incident response simulations: Conducting hands-on exercises where the team responds to simulated cyber attack situations can help better prepare staff for real incidents.

The above actions are essential to ensure the safety and reliability of autonomous vehicle systems. Effective implementation of sound practices from design to validation not only protects against technical errors and malicious attacks, but also builds public confidence in this emerging technology. As autonomous vehicles continue to develop and integrate into our societies, ensuring their safety will be critical to their widespread adoption.

Software Testing and Validation

Rigorous testing is essential to ensure that the software operates properly under different operating conditions prior to actual deployment:

- **Unit tests:** Validate individual functions within code ensuring correct isolated operation before overall integration with other components; this facilitates early detection of bugs to prevent more serious problems during the later stages of development and finalized implementation.
- **Functional and integrated testing:** More extensive evaluations ensure that the interactions between modules operate as expected under realistic representative simulated conditions; this encompasses both physical and virtual

testing in which complex situations commonly encountered during day-to-day operation are simulated.

These evaluations should be carried out on an ongoing basis throughout the software lifecycle; this ensures not only quality but also overall reliability as new and additional functionality is added over time.

Risk Management & Incident Response

Proper risk management is crucial, given the high level of involvement in public safety. This involves identifying potential threats and establishing clear incident response protocols:

Continuous assessment of potential risks: This encompasses the detection of cyber risks and internal technical errors that could jeopardize the security of the system.

Establishment of clear procedures: It is necessary to establish protocols for an immediate reaction to incidents, ensuring the reduction of collateral damage and reducing the overall adverse effect on the end users involved.

Development of recovery strategies after an incident: It is crucial to ensure rapid recovery of normal functionality after any negative event, without compromising security during the recovery process.

Establishment of minimum standards: It is essential to establish minimum standards to ensure the quality and proper functioning of the software systems used in autonomous vehicles.

Creation of clear regulatory frameworks: It is necessary to establish regulatory frameworks that establish obligations and supervise compliance with the regulations in force, thus ensuring the safeguarding of end users.

Collaboration between governments and regulators: Collaboration between government agencies, regulators and the private sector is a key factor in the development of the company's business model.

essential to promote the advancement of best practices, ensuring safety and building public confidence in the large-scale adoption of these new means of mobility.

Implementation of regular audits: It is essential to conduct regular audits to confirm compliance with regulations and to ensure ongoing compliance with established standards.

These actions are essential to ensure that autonomous vehicles operate safely and effectively in a regulated context (Bhatia et al., 2021)._Thus ensuring that the risks associated with their operation are adequately managed (Gonzalez et al., 2021).

The Future of Autonomous Vehicle Safety

The future of security in the context of autonomous vehicles promises to be exciting, but also challenging. As the underlying technologies continue to evolve, emerging concerns related to the protection of the end users involved such as the following must also be addressed:

Development of innovative solutions: It is vital to increase resilience to cyber attacks by ensuring the integrity of critical operating systems used in vehicles (Bhatia et al., 2021). This entails the implementation of proactive actions capable of identifying and reducing threats before they cause significant damage.

Implementation of advanced artificial intelligence: The ability to identify irregularities in irregular behavior in the communications network is essential to detect potential threats before they can effectively materialize (Zhou et al., 2020). This requires the implementation of advanced algorithms that examine patterns and behaviors in real time.

Fostering public-private collaboration: Sharing data and best practices among governments, regulators and private industries is essential to ensure appropriate preparedness for potential future incidents associated with this new mode of transportation (González et al., 2021). Collaboration between sectors can promote the creation of standards and protocols that increase overall safety.

Incorporation of continuous end-user feedback: Providing prompt and effective responses to submitted safety-related concerns is crucial to preserve public confidence in the widespread adoption of these new means of modern mobility (Chen et al., 2019). Proactive feedback can help detect areas requiring improvement and modify policies and technologies accordingly.

These strategies are fundamental to building a safe and reliable future for autonomous vehicles, ensuring that the risks associated with their operation are properly managed.

Social Impact and Future of Mobility

The advent of autonomous vehicles is transforming cities, the economy and the environment, while posing significant ethical and legal challenges. This technology promises to redesign urban infrastructure, create new job opportunities and foster sustainability, while displacing traditional roles and demanding clear regulations for responsible implementation.

In this section, we will explain how autonomous driving impacts infrastructure, employment, the economy and sustainability, reflecting on its potential to improve quality of life and connect communities more efficiently and sustainably.

Changes in Urban Infrastructure

The massive arrival of autonomous vehicles will radically transform existing urban infrastructure; this includes:

Redesign of streets and intersections: Traffic flows will be improved by taking into account the particular demands linked to the booming new modes of transportation. This restructuring is crucial to promote the incorporation of autonomous vehicles in the urban context, increasing traffic safety and efficiency (González et al., 2021).

Implementation of fast charging stations: In the future, strategically located accessible charging stations will be installed, ensuring the availability of energy required for the constant operation of these new means of modern mobility. Proper organization of these stations is essential to promote the widespread adoption of electric and autonomous vehicles (Bhatia et al., 2021).

Development of friendly public spaces: Places will be built where cyclists and pedestrians can coexist harmoniously with automated electric vehicles, thus promoting sustainable and healthy lifestyles in the community. This perspective aims not only to increase air quality, but also to promote more intense social interaction in urban environments (Zhou et al., 2020).

Integration of intelligent technologies: Smart technologies will be implemented in the current infrastructure, facilitating efficient communication between automobiles and urban infrastructure. This will increase the overall performance of the urban transport system, promoting more efficient traffic management and decreasing travel times (Chen et al., 2019).

These modifications are essential to equip cities for the future of mobility, ensuring that infrastructures are adjusted to the new technological realities.

Effects on Employment and the Economy

The economic impact linked to the massive implementation of this new means of transport will be considerable; this encompasses as many new possibilities as there are challenges in the current labor market:

Job creation: Jobs will be created linked to the advancement and preservation of the technology that supports these new means of contemporary mobility. This includes jobs for engineers, developers, and cybersecurity specialists, among others (Bhatia et al., 2021). The need for competent professionals in these fields will increase as the autonomous vehicle industry develops.

Displacement of traditional jobs: Automation and the implementation of independent vehicles could replace traditional jobs linked to manual driving. This will trigger the need to train the current workforce to adjust to the new demands and competencies needed in future emerging industries linked to this field (González et al., 2021). It is crucial to establish training programs that will help employees in their transition to more technical and specialized positions.

Changes in consumption patterns: The decrease in transportation costs will promote development in nearby economic areas, such as logistics, distribution and e-commerce services. These modifications will foster an economic boost across the board, which could lead to an increase in labor employment and improvements in market efficiency (Zhou et al., 2020).

These factors underscore the relevance of training both workers and infrastructure to maximize the financial benefits from the deployment of autonomous vehicles.

Ethical and Legal Aspects

Legal aspects related to civil liability are fundamental in the context of autonomous vehicles. As this technology advances, it becomes necessary to establish new regulations defining who assumes liability in the event of an accident. This involves considering both the producer of the vehicle and the owner. The complexity of this situation lies in the nature of the operation of autonomous vehicles, which operate using algorithms and artificial intelligence systems, raising questions about liability in situations where a critical decision must be made by the vehicle.

Civil liability in the field of autonomous vehicles must be reevaluated to adapt to the new realities. Traditionally, liability rested with the driver; however, in the case of an autonomous vehicle, it is necessary to determine whether the fault lies with the software manufacturer, the vehicle owner or even the artificial intelligence system itself. The European Union has begun to address these issues by revising existing laws and introducing directives that seek to improve the protection of victims in accidents involving automated vehicles (Michel, 2020).

In addition, clear guidelines must be established on how to handle situations where an autonomous vehicle must make morally complicated decisions during an impending accident. For example, if a vehicle must choose between saving its occupants or a group of pedestrians, the question arises: who is responsible for that decision? This ethical dilemma not only affects vehicle programming, but also has significant legal implications (Bustamante Donas, 2022).

Sustainability and Environment

Autonomous vehicles have the potential to contribute significantly to more sustainable mobility. Their implementation can transform the dynamics of urban transport, optimizing traffic and thereby reducing overall emissions. This translates into a considerable reduction in energy consumption associated with transportation, especially in congested urban areas. The ability of autonomous vehicles to communicate with each other and with the road infrastructure enables more efficient traffic management, which can minimize congestion and improve vehicular flow (Fagnant & Kockelman, 2015).

The effectiveness of autonomous vehicles in improving sustainability depends largely on the public policies that accompany them. It is critical that governments establish regulations that promote the adoption of clean technologies and incentivize car sharing. Policies that encourage the electrification of transportation are also crucial, as electric vehicles combined with autonomous capabilities can further reduce carbon emissions (Bösch et al., 2018).

It is important to consider the rebound effects associated with the adoption of autonomous vehicles. These effects may manifest as an increase in overall vehicle use due to convenience, which could offset some of the expected emission reductions (Feng et al., 2020). Therefore it is essential that implementation strategies include measures to mitigate these effects.

Current studies indicate that if public policies favoring these new vehicle models are correctly implemented, a significant decrease in the ecological impact linked to private transport could be achieved (Fagnant & Kockelman, 2015).

Future Trends

The future promises continued innovation, where we expect to see greater integration between different modes of transport; for example, intermodal systems where users can easily switch between traditional public transport and services based on autonomous vehicles.

As we move into a future where these vehicles are increasingly common, it is essential to address both their benefits and inherent challenges through continued research and responsible implementation.

Conclusions

The development of autonomous vehicles marked a milestone in the history of technology, challenging the limits of human knowledge and transforming our interaction with the world. This book explored in depth the fundamentals, technologies, and challenges that made this revolution possible, making it clear that these advances were much more than technical achievements; they represented a profound change in the way we conceive of mobility and our relationship with technology.

The integration of artificial intelligence, perception systems and software architecture enabled autonomous vehicles to optimize mobility and reduce human error in transportation. However, these advances also raised ethical questions and social challenges that demanded a responsible response from those leading their development. Reflecting on these challenges and the proposed solutions was key to understanding how technological progress was balanced with the need to protect people's rights and privacy.

The social and economic impact of autonomous vehicles was remarkable. They changed and will continue to change urban infrastructure, generated new job opportunities in technology sectors, and redefined the economy of entire cities. But these achievements also brought with them the need to transform the skills of the workforce and to rethink the design of our cities to adapt them to more efficient and sustainable mobility.

With this work, a fundamental chapter in the understanding of how humanity was able to shape the future with responsibility and creativity was closed. The reader is invited to reflect on how, at the time, these decisions shaped not only a safer and more efficient transportation system, but also a society that was more connected and aware of the ethical challenges that accompanied progress.

REFERENCES

Advancement of Machine Learning and its Applications (n.d.). *Semantic Scholar.* https://www.semanticscholar.org/paper/15d4f93b210a354f6ca6f78339c2 bdc6739 .2e9af

Anderson, J., Kalra, N., Stanley, K., & Sorensen, P. (2016). *Vehicular autonomy: Evaluation and implications.* RAND Corporation.

Bar-Shalom, Y., Li, X.-R., & Kirubarajan, T. (2001). *Estimation with Applications to Tracking and Navigation.* Wiley-Interscience.

Bhatia, P., Jain, S., & Sharma, A.K. (2020). Review of radar technology for automotive applications: Current trends and future prospects in vehicle automation systems and safety measures. *IEEE Transactions on Intelligent Transportation Systems*, *21(3),* 1126-1137.

Bhatia, P., Kaur, S., & Kumar, A. (2021). Recent advances in motion and behavior planning techniques for software architecture of autonomous vehicles: A state-of- the-art survey. *Journal of Autonomous Vehicles and Systems*, *1*(1), 1-15. https://doi.org/10.1109/JAVS.2021.3050489. https://doi.org/10.1109/JAVS.2021.3050489

Bösch, P., Becker, H., Becker, J., & von der Gracht, H. A. (2018). Autonomous vehicles: The impact on urban mobility and the environment. *Transportation Research Part D: Transport and Environment*, 61, 1-17.

Bustamante Donas, J. (2022). Ethical dilemmas of autonomous vehicles: Ethical responsibility, risk analysis and decision making. *Arguments of Technical Reason.*

Chen, W., Zhang, Y., & Wang, H. (2019). Distributed architecture for autonomous vehicles based on multi-agent systems. IEEE Transactions on Intelligent Transportation Systems, 20(6), 2210-2220.

Crispin, L., & Gregory, J. (2009). *Agile Testing: A Practical Guide for Testers and Agile Teams.* Addison-Wesley.

Doherty, P., O'Neill, M., & McCarthy, J.P. (2017). Sensor fusion for autonomous vehicles using deep learning methods: A review of recent advances and future challenges in intelligent transportation systems and road safety applications*. *IEEE Transactions on Intelligent Transportation Systems, 18*(8), 2150-2164.

Dosovitskiy, A., Roscher, R., & Becker, M. (2017). CARLA: An open urban driving simulator. *arXiv preprint arXiv:1711.03938.*

Fagnant, D. J., & Kockelman, K. M. (2015). Preparing a nation for autonomous vehicles: opportunities and challenges. *Transportation Research Part A: Policy and Practice, 77,* 167-181. https:ZZdoi.org/10.1016Zj.tra.2015.04.003.

Feng, T., Wang, Z., & Chen, Y. (2020). Exploring the rebound effect of autonomous vehicles on energy consumption and emissions in urban areas. *Sustainable Cities and Society*, 53, 101883.

Fitzgerald, B., & Stol, K.J. (2017). Continuous Software Engineering: A Roadmap and Agenda. *Journal of Systems and Software, 123,* 176-189.

Gónzalez, J., Pérez-Cruz, F., & Bañares-Alcántara, R. (2021). A review of perception systems for autonomous vehicles: Challenges and future directions. *Sensors*, *21*(8), 2757. https:ZZdoi.org/10.3390Zs21082757.

Goodall, N. J. (2014). Ethics of automated machines and vehicles. In *Road Vehicle Automation* (pp. 93-102). Springer.

Goodfellow I., Bengio Y., & Courville A.(2016). *Deep Learning.* MIT Press.

ISOZSAE 21434. (2021). *Road vehicles - Cybersecurity engineering.* International Organization for Standardization.

Kamin, D.A. (2017). Exploring Security, Privacy, and Reliability Strategies to Enable the Adoption of IoT. *Doctoral Study,* Walden University.

Kumar, A., Gupta, R., & Singh, S. (2020). Model predictive control for autonomous vehicles: A review and future directions. *IEEE Transactions on Intelligent Transportation Systems*, *21*(6), 2539-2554.

Larman, C., & Basili, V. R. (2003). Iterative and incremental development: A brief history. *Computer, 36*(6), 47-56.

LeCun Y., Bottou L., Bengio Y., & Haffner P.(2015). Gradient-based learning applied to document recognition*. *Proceedings of the IEEE, 86*(11), pp.-2278-2324.

Litman, T. (2020). *Transport policy and the environment.* Victoria Transport Policy Institute.

Meyer, M., Huber, M., & Schmid, K. (2020). Agile Development in the Automotive Industry: Challenges and Opportunities for Product Development Processes. In *Proceedings of the International Conference on Agile Software Development* (pp. 1-15).

Navarro Michel, M. (2020). *The application of traffic accident regulations to those caused by automated and autonomous vehicles.* Semanticscholar. https://www.semanticscholar.org/paper/24e1ac657fd8512b94a5b7cd5a1 e138b5 .78cc8bb

SAE International. (2018). *Taxonomy and definitions of terms related to driving automation systems for on-road motor vehicles* (SAE J3016 201806). https://www.sae.org/standards/content/j3016_201806/

Schwaber, K., & Sutherland, J. (2017). The Scrum Guide: The Definitive Guide to Scrum: The Rules of the Game

Tesla Inc (2021). *Tesla's Autopilot.* https://www.tesla.com/autopilot

VeRA (2020). Vehicles Risk Analysis Methodology for Autonomous Vehicles. Retrieved from https://www.semanticscholar.org/paper/d9e441741d5c3aff2dde27e543cc 07f499e c03fa

Waymo LLC (2021). *Waymo's self-driving technology.* https://waymo.com/technology/

Welch G., & Bishop G.(1995). An introduction to the Kalman filter*. In *SIGGRAPH*

Course Notes.
Zhou, Y., Chen, J., & Zhang, Z. (2020). Deep learning for autonomous driving: A review of recent advances and challenges ahead. *IEEE Transactions on Intelligent Transportation Systems, 22(5),* 2764-2778. https://doi.org/10.1109/TITS.2020.2970512.

Printed by Books on Demand GmbH, Norderstedt / Germany